Sweet Mystery of Life
A Judeo-Christian Exegesis

June Raleigh

Table of Contents

Foreword...v

Part I   God is Life... 1

Chapter 1   Spirit and Water 2

Chapter 2   The Blood.................................. 11

Chapter 3   Influence of Man....................... 16

Part II   What God is Like..............................23

Chapter 4   Geology and History 24

Chapter 5   Moses and the Rock..................... 31

Chapter 6   Judaism and Christianity........... 40

Part III   For the Love of Jesus 51

Chapter 7   Mary Magdalene........................ 52

Chapter 8   The Virgin Mary 56

Chapter 9   In Defense of Women in
            the Church................................ 61

Chapter 10   God's Message 66

About the Author...................................... 72

Foreword

It's all about people, the folks in this blessed country and all those who search for God in the world. This book is written for you, and anyone who seeks answers to what they are doing here and the meaning of their life. At some point most of us want answers about love, death and everything in between. Take this message to heart if you care about strengthening your soul, and not just exist sitting around waiting for time to pass. It passes all right, and quickly. All your possessions will not go with you where you are going. The only thing you truly have is your soul, and that is the thing you can make stronger to prepare for the day you leave this earthly life.

"I gave in, and admitted God was God." -C.S. Lewis

Once you ask God to come into your life and stop willfully sinning you will get a gift. Everyone's gift is different, but what is true for all is once God blesses you, you will know what to do with your life. In this way you will do what you love, not just the labor of the day. Of course, choosing inaction makes you subject to other people making decisions about your life for you. This is called fate, and fate tosses your life to the wind when the bad times hit. Guard against this by making your own choices and control your destiny, so that you may prevail in the day of trouble. Persistence wins in the end: "I again saw under the sun that the race is not to the swift, and the battle is not to the warriors, and neither is bread to the wise, nor wealth to the discerning, nor favor to the skillful; for time and chance overtake them all" (Ecclesiastes 9:11) Even when your heart is broken, with fortitude you will not kill yourself; remember how God feels about that: "A broken and contrite heart, O God, you will not despise" (Psalm 51:17)

The ancient Hebrews who developed the culture of Rabbinic Judaism, which resulted in the Bible gave us the greatest contribution to western civilization, and to the world in general. Many countries to this day are designed on the basis of the Ten Commandments. It is important for the devout believer in Messiah Jesus to understand how the New Testament is derived from the Old Testament. The catalyst that unites Judaism to Christianity is God's completion of a grand plan to save souls onto Himself. He did this by first presenting rules to live by to the Hebrews, then finished what He started by offering a path to the afterlife through salvation in the New Testament. May this book be a useful exegesis and a blessing to you.

Part I

God is Life

Chapter 1
Spirit and Water

Life is precious, and the beauty of the earth is exquisite. When you realize how quickly it all goes by and begin to reflect on your own mortality, you remember those excellent times when you were at your happiest; great memories of when you experienced the joy of life whether with those you loved, at places you lived and visited, or both together. You might be wondering what it all means- your life exists in the totality of being in consciousness. You are part of the human quest for meaning, and the life force of the earth.

Opportunities change as you grow, and things can be very different from the way they were. You get wiser and learn to savor every moment you share love with people and animals as well. It takes a lifetime to understand the importance of love- life has no meaning without it. After you obtain the basic necessities the real essence of life is love. A younger person may not quite understand all this yet, but if they show reverence for God they are cultivating a good soul and have a chance to obtain salvation. In fact the younger you are when you search for God the better.

Let us ponder the sacred dimensions of interaction between God and mankind by aligning human everyday life and spiritual phenomenon with scientific truth, with the suggestion that God designed them together. Religion and science are occurring simultaneously, and when you see where they intersect you begin to realize the profundity of life.

One dimension is time and why does it exist in two different measurements, first here during our lives, then in the eternal? These are two different definitions of time: what we know is a twenty-four hour clock, but what about God's eternal afterlife? Psalm 90:4 tells us "For a thousand years in your sight are like yesterday when it passes by, or as a watch in the night." So one day or less in heaven is equivalent to a thousand years in our flesh lives. Mathematically there is no exactitude to the quotient, because one hour in heaven is approximately 41.6 years on earth. Even so the product is still not precise, in that 41.6 times twenty-four hours equals 998.4 years, just under a thousand. Time for us may be approximate and fluid, as illustrated by the insertion of a leap year day (February 29th) every four years, and scientifically it is entirely relative to motion and gravity. Amazingly, the Bible is compatible with modern astrophysics: we know when in space astronauts experience time slower than on earth- how did ancient Judaism know this?

The prevailing belief of the ancient Hebrews about the thousand years being like one day to God was that it's simply an idiom, not to be taken literally as a time measurement. The belief is that God is in an eternal timeless dimension, apart from our linear dimension of time.

Since the moon moves away from the earth at 3.78 centimeters (just under an inch and a half) per year time is adjusted by adding a second to our clock every 1.8

years, or 26 times in the last 48 years. This gravitational pull slows the earth's rotation ever so slightly, causing the day to become longer. It seems like your life span is shorter but it's not. Ancients in the Bible lived over a hundred years; perhaps this gives a partial, minute (no pun intended) explanation of how.

So far in physics time is proven to be linear, always going forward. There is no evidence that we can go backward into the past, although theoretically you can see the past if light year speed and location variables are employed. 1. You may see it but you can't change it because once it's happened it's impression is there forever; that is if space extends forever. The photons of light keep traveling through space like a movie of your life.

Looking back nearly four thousand years, within the everyday life of ancient Judaism there were three watches, from sunset to 10PM (Lamentations 2:19), the middle watch from 10PM to 2AM (Judges 7:19), then the morning watch from 2AM until sunrise (Exodus 14:24). (During Roman rule this was expanded into four watches with obviously shorter time spans). For Israel, the first is the beginning of the next day, a time for reaching out to the Lord, the second is ripe for a surprise attack against your enemy, and the third is when God himself appears or takes action. "Weeping may last for the night, but a shout of joy comes in the morning." (Psalm 30:5) When the sun rises in the east it illuminates our day and the watches are over. Serenity and tranquility while lying in a green pasture in the sun is different from satisfaction and safety while sleeping at night. There are different types of peace such as security, simple satisfaction and deep contentment. Life used to be much simpler, but now in this unforgiving age of technology it's much more complicated. One thing has not changed and that is we all seek peace. It is found in the peace God gives us. Notably, the embodiment of peace, and much more is found in the divine Hebrew word "shalom." However, there are no Latin, Greek or English translations for this most important and expressive word.

God gives us time to grow love and virtue into our lives so we can understand him and be more like him. Eternal life is an alliance of love between God and man. He is in the business of saving souls and that is befitting since He is a spirit. "But an hour is coming, and now is, when the true worshippers will worship the Father in spirit and truth; for such people the Father seeks to be His worshippers. God is spirit, and those who worship Him must worship in spirit and truth." (John 4:23-24)

Metaphysically speaking, you are who exists in time on earth right now as a witness to what life is about- all people and animals are part of the blood and water of flesh. The experience of your life makes an indelible print of truth onto your soul- it is what it is. On earth there is **"The Spirit, the water and the blood and the three are for one thing."** (1 John 5:8) This is a tremendous passage and one that gives us the existential components of our being, a central message of life. You are automatically part of the water and blood because you are here, but it is a deliberate step toward eternity to gain the holy spirit into your life.

The Holy Spirit, who is within God and yet is also here was in heaven when Jesus was on earth. When Jesus was facing his death he told his disciples he would send them help: "And I will ask the Father, and He will give you another Advocate to be with you forever- the spirit of truth." (John 14:16,17) Jesus in effect switched places with the Holy Spirit or Ghost, so believers could be comforted when Jesus went back to the Father after his job was completed. They worked in tandem: "Unless I go away the Advocate will not come to you; but if I go I will send Him to you." (John 16:7) Initially the Advocate was a separate person from Jesus because he's saying he has to go and get him. This could be part of the reason Jesus had to come to earth and die, so the Holy Spirit could be brought into our world to more fully save our souls. This also explains why Jesus cried out in his last moments "My God, my God why have you forsaken me?"(Matthew 27:46) Jesus spoke Aramaic but the word he used was "paracletos" (Greek) which means advocate, helper, bringer of glad tidings, and someone who brings hope: these are names for the Holy Spirit, who came after Jesus ascended and now bears witness. Clearly the Holy Spirit is truth itself: "And it is the Spirit who bears witness, because the Spirit is the truth." (1 John 5:7)

The Holy Spirit in Hebrew is "Ruach Ha-Kodesh." Ruach means "breath" or "wind," and Hakodesh means "holiness." Isaiah 11:2 tells us the gifts of the spirit are "wisdom and understanding, the spirit of counsel and strength, the spirit of knowledge and the fear of the Lord." Ruach Ha-Kodesh has no gender, although "Ruach" is a feminine noun; then again, Jesus' description of the Holy Spirit "paracletos" (Greek) is masculine. God may be nonsexual, or above gender definition. 2. Maimonides, known as "Rambam" in his theory of negative attributes suggests that we may only describe God in terms of what God is not.

God's spirit appears from the beginning, where in Genesis 1:2 "the spirit of God was hovering over the face of the waters." The Rabbinic understanding of God is that He is monotheistic, as opposed to the Christian tripartate God that is three persons in one. Judaism believes the Holy Spirit for example, is an attribute of God, but not another separate person within God. The only singular being that exists is God, as is recited in the Shema in Deuteronomy 6:4: "Hear Israel! The Lord is our God, the Lord is one!" In the first verse of the Bible He is called Elohim, which is not a name, but in Hebrew means "of majesty." To elucidate Elohim, a masculine, plural noun, the truth is God is one but many. The Christian concept of the Trinity comes from this. If you put the two views of Judaism and Christianity together Jesus the Messiah brings God closer to humanity as an attribute within the one God.

Jesus described the body as a "sanctuary," a temple where the Holy Spirit can reside. (John 2:19-21) He also said you must be born again from above, "born of water and spirit" to "enter into the kingdom of God." (John 3:5)

Water is the voice, the sound of the Word of God: "I saw the glory of the God of Israel coming from the east. His voice was like the roar of many waters and the earth shone with his glory." (Ezekiel 43:2) Again, in Revelation 1:15: "His voice was like the roar of many waters." Our earth is 3/4 water, way too much to have accumulated from chance asteroids containing water that bombarded the earth, although many did. We know that early in our sun's formation it emitted a solar cloud nebula of gases and hydrogen. Through a chemical system called iron hydride, hydrogen absorbed into earth's core through its attraction to iron and is now continuously released. Plants make oxygen, and since water is 2 hydrogen molecules plus 1 oxygen molecule, over a vast amount of time we have the accumulated production of water, a natural causation of God's handiwork. Earth makes its own water; here again God and science are one. This is a chemical paradigm of how oxygen, hydrogen and iron behave under the mantle of the earth, whereas closer to the surface oxygen increasingly releases and disperses.

When God created the earth there was no rainwater falling from the sky. Instead, water came from the earth's crust. Genesis 2:6 states: "But a flow used to rise from the earth and water the whole face of the ground." The ground was kept moist from water pressure coming up onto the earth's surface; streams and rivers were thus formed, and these branched out extensively.

Smoke from volcanic fire drifted up between ten to thirty miles above the earth in the area called the stratosphere and together with nitrogen and oxygen formed the protective ozone layer, which blocks harmful radiation. The ozone layer began about 600 million years ago and today it's healing itself despite mankind's damage to it with chlorofluorocarbons.

Has God done this before? Probably. He got a giant ball of iron just the size He wanted and placed it 92.96 million miles from the young sun to achieve the desired temperature zone. He wants to create life and knows very well the formula to do so. Astrophysicists say the chunk of iron that is the earth's core broke off from the sun about 4.6 billion years ago, and the moon broke off from the earth, but it is a marvel that the earth is exactly at the right distance from the sun to create a temperate climate zone just right for life.

If anyone has ever hiked into mountains they've probably seen clouds forming above the trees: these clouds of water vapor evaporate from the trees, sort of like when you sweat water comes out of your pores. The trees save us from the ravages of CO2; the more trees the better. Since earth is a closed ecosystem the atmosphere holds the water within so it cannot escape into space- it is simply recycled. Of course we all learned in the fifth grade about photosynthesis, how plants take in carbon dioxide and convert it to oxygen, a process that supplies the essential element for all life on the planet. When we exhale carbon dioxide feeds plants, so our existence and all life that breathes is part of an interchangeable cycle of planetary life. All life must breathe; what a coincidence that the acronym of God's holy name YHWH sounds like breath itself when pronounced.

The firmament, or heaven was created by separating the waters above from below. Heaven has rivers of water for the garden of Eden which was tended by Adam and Eve. The firmament water is fresh, not salty as it is purified from what its composition was below. The firmament, or in Hebrew "raqia" emphasizes the solidity and purity of this vault that God created. He separated the waters and put in the expanse, which is the sky between them. Historically in context with the Babylonian and Egyptian definitions of the firmament it was solid. Genesis 1:6 describes the creation of the firmament on the second day: "an expanse in the midst of the waters". The most compatible description with Genesis of the firmament is found in Ezekiel 1:22: "the awesome gleam of crystal" and the combination of the two is found in Revelation 4:6 and Revelation 15:2, where in both passages we see the throne of God atop the "sea of glass."

By the time God created the plants it was the third day. As explained above, one day for Him in heaven is equivalent to a thousand years. If we take that formula and apply it to an earth day wherein God is working creating earth, it took Him six thousand years to basically structure earth's environment. If you base your beliefs on empirical evidence you can say God incorporated science in His actions of creation between 3.8 to 4.5 billion years ago, when astrophysicists say was about the time water first existed on earth. Opinions of when the earth first existed vary considerably among scientists and others, but the scientific, chemical elemental processes noted above are not debatable.

1.   *Nytimes.com/2006/05/16/science/16ligh.html*

2.   *Bruce K. Waltke and M. O'Connor, An Introduction to Biblical Hebrew Syntax, 1993.*

Chapter 2

The Blood

Mankind is the blood; all must die for the remission of sins. However, in your death there is the possibility of atonement through Jesus' blood, which yields forgiveness and redemption. "For the life of the flesh is in the blood, and I have given it to you on the altar to make atonement for your souls; for it is the blood by reason of the life that makes atonement." (Leviticus 17:11) Death of the blood began with the sin of Adam and Eve because of disobedience to God. The power of the blood to restore life reached its apex in the blood of the Son of God Jesus because of his sacrifice.

Human blood is about 4% salt, nearly the same amount of salt as the oceans (3.5%). Of course some ocean areas are more or less saltier where fresh water merges into the seas or not. Interesting that all the elements in our blood are also present in sea water, although not in exact configurations with each other. When Jesus said "You are the salt of the earth" in Matthew 5:13 he wasn't kidding. Salt is important to God, as He makes a point of emphasizing it several times in the Bible. It stands for preservation of meat, purification of water, and a token of trust, such as the "covenant of salt" between the Lord and King David, where God promised David he and his descendants kingship over Israel forever. (2 Chronicles 13:5) Here, God establishes David's bloodline, and the salt therein is like a DNA print of it. Although David followed everything God wanted through battles and governance of His people and he was "a man after His own heart" (1 Samuel 13:14) he lost God's trust to the degree that God would not allow David to build a temple for Him. Instead God chose to have David's son Solomon build the temple. God explained to David this was "because you are a man of war and have shed blood." (1 Chronicles 28:3) God's decision was probably a result of the actions David took against Uriah the Hittite when David had him killed by putting Uriah on the front line of battle while committing adultery with Uriah's wife; despicable actions in the sight of the Lord. The shedding of blood matters.

Another reason God stopped David from building a temple is found in the Talmud (Devarim 11:24), 1. when David took it upon himself to conquer two Syrian cities, Aram-Naharim and Aram-Tzovah. Rabbis Rashi and Tosafot state David did not act in accordance with the Torah because he was supposed to first drive out the Jebusites near Jerusalem before setting out to take other cities. Later on David did conquer the Jebusites, and that area became The City of David (2 Samuel 5:6).

Blood is very necessary for your life as more than two thirds of the iron in your body is in your red blood cells, and your blood is from 38 to 48 percent iron. The iron in your blood's hemoglobin carries oxygen throughout your body. Is it any wonder the earth's core is iron and that it is compatible with oxygen? Male bodies store iron far more efficiently than females do and can get it from food, but females must take iron supplements.

You are part of the process of the witnesses because you are fundamentally of the blood, and therefore part of the life force of earth. You have an inheritance in the afterlife if you choose to recognize how to obtain it, by having faith and following God's laws. It's not overly complicated, but it does take time to understand. "Anyone who has set aside the Law of Moses dies without mercy on the testimony of two or three witnesses." (Hebrews 10:28). Lawlessness is condemned. Some will say but Jesus saves by grace, not by law- yes, it's God's choice to bestow grace that leads to faith in you, but keeping His precepts are important: Jesus said "Do not think that I have come to abolish the Law or the Prophets; I did not come to abolish but to fulfill." (Matthew 5:17) So Jesus is the nexus between the two forces of law and faith. Naturally, corollaries to the law are good works, something you do after you are saved. For example, it would be impossible for a saved person to delight in actions that go against the Ten Commandments. The central, important point is that only through the atonement by Jesus' blood are you offered salvation, and everything about this mystery revolves around that. "This is the one who came by water and blood, Jesus Christ; not in the water only, but in the water and in the blood." (1 John 5:6)

When Jesus stood on the mountain of Transfiguration (Matthew 17) he was flanked by Moses the lawgiver on one side, and the prophet representative Elijah on his other side. In centuries past, first came the law, then came several prophets who predicted Jesus' coming. Moses recorded the word and will of God, and Elijah stood for the blood of the prophets, many who were slain for their predictions of the coming Messiah, as well as all of the prophets that God gave to each king of Israel as a helpmate. Jesus is the voice of many waters, the Word that incarnated as God's promised Messiah the Christ, and he is also cross referenced in Hebrews 10:29 as the final sacrificed blood for the covenant that was first presented by Moses in Exodus 24:8. However, it is important to note that in Rabbinic Judaism the tradition is that the Word manifested as the Torah, inasmuch as orthodox Jews do not accept Christ unless they are Messianic Jews. For Christians the sacrifice of the savior who is the blood and the water together with the follow up of the Holy Spirit are for one thing: to glorify God through the salvation of souls, especially of mankind- this is God's design operating in a closed ecosystem of life, unfolding His desire to save unto Himself many of the souls of the earth. Giving you eternal life is how He expresses His gracious love.

When you die your body is done, but your soul is not necessarily recycled back to the same environment of earth. That real essence of you that had formerly borrowed your flesh body is changed into the best your spirit can be, hopefully destined for heaven. We have already seen that technically there is no going back, only moving forward, as time is linear. Sorry, there is no reincarnation in the Bible as explained "and inasmuch as it is appointed for men to die once and after this comes judgement" (Hebrews 9:27). Like a chrysalis that metamorphosizes into a butterfly you are made to "put on the new self, which in the likeness of God has been created in righteousness and holiness of the truth." (Ephesians 4:24)

Your soul has more than one life and will certainly live on. The question is where your soul will end up. That real essence of you that is your soul is called in Hebrew "Nephesh." In Genesis 2:7 God breathed life "chayyim" into man, making him a living soul. The suffix "im" added to the end of a word makes it plural. Nephesh chayyim literally means "soul lives" which tells us that the human soul has more than one life. That is why we occasionally see ghosts cluttering the earth. Heaven is not automatic, and only the truly righteous who effectually serve God are rewarded with ascension out of the physical earth into a spiritual existence with Him.

There is no clear direction in the Old Testament on how to get eternal life for your soul, but there is a path in the New Testament, which is secured by faith in the Messiah Jesus. Many Israelites at the time of King David had thought perhaps he was the Messiah, but that was not so; however David is considered to have been the greatest king of the Israelites, who was dedicated to following God's laws and chosen to have the Messiah succeed from his lineage. The theophany God provided connects the Old to the New Testaments from what was predicted to what manifested as Jesus Christ as the Messiah.

1. *Sefaria: Sifrei Deuteronomy 51:2, Devarim 51, translated by Rabbi Shraga Silverstein, Wikisource.org, public domain*

Chapter 3

Influence of Man

Many versions of Bibles read "For there are three that bear witness in heaven, the Father, the Word, and the Holy Spirit, and these three are one." (1 John 5:8). In the early 1700's scientist Sir Isaac Newton found that in the original Greek text, which is the language the first New Testament was written in, this passage about the Trinity doctrine was not there. 1. He contends that this biblical writing was added in by a Catholic writer Cardinal Ximenes in 1515. Lending credibility to Newton's finding, many Bibles used today in several Eastern nations including Egypt, Ethiopia, Syria, Armenia, Arabia and others still do not and never have had this passage. The passage from 1 John 5:8 we are expounding on about the three on earth was always there, but not so the one about the three in heaven.

Additionally, the 1 Timothy 3:16 passage "And by common confession great is the mystery of godliness: He who was revealed in the flesh..." was also edited by Catholics. The original said: "Great is the mystery of Godliness, which was manifested in the flesh." Newton's findings were finally applied in the 1800's, as these opinions were not voiced during Newton's lifetime because of the penalty of death to those who did not accept Catholic doctrine. 2. For example, an eighteen-year old named Thomas Aikenhead was hanged for disputing the trinity in 1697, and Newton's friend William Whiston lost his position at Cambridge in 1711. The meaning that can be derived by the addition of the words "common confession," refers to the Catholic practice of confession of your sins to a priest from behind a veiled closet. However, confession in the Bible does not require a veil and is instead simultaneous with actual repentance and baptism, as advocated by John the Baptist. Moreover, the word "common" trivializes and is non-sequitur to the word "mystery."

So there were the Catholics adding words to the New Testament, and there are many examples of it, one of which I just gave you. Yet I have been told countless times by churchgoers that "the Bible is the inerrant word of God" (this quote is not biblical by the way) and I am to accept whatever it says without doubt. When I challenge this argument and cite changes made centuries after Christ's resurrection, or passages added by scribes from the secretary pool in Ephesus pretending to be Paul after his death as another type of example, the rebuttal is usually "it doesn't matter, it made it in there, therefore it's part of God's word." If you apply logical reasoning to this, then anything goes because once it happens it's a done deal. It doesn't take a rocket scientist to see that this is not a rational way to think. To use an analogy, suppose your brother in-law comes to stay with you for a few days, and you give him the room downstairs. However, he really likes it at your house and after a week he attends his community sponsored mechanic classes. Soon he brings your teenage son to these mechanic meetings with him, and your son changes his aspiration from wanting to be an electrician to now becoming a car mechanic, due to the influence of your brother in-law. Here, changes are occurring in degrees that are not immediately perceptible. Cause and effect are not always closely related in time. Since he offers to pay two hundred dollars a month for the room, your wife advocates for him, and now you have a permanent change to your household. As time passes your in-law makes more subtle changes that you learn to live with but were never part of your original household plan. If you are able to think for yourself you are not going to say it doesn't matter because it happened. No, a particular set of conditions affecting how things occur does matter. God inspired the Bible, but man wrote it, and apparently a certain group more than others kept on fashioning the New Testament to their advantage for well over a thousand years after the death of Christ. Shall we make an exception and say all who wrote in the Bible were inspired by God? All right, but then why did the Council of Nicea in 325 AD take so much out of the canon?

When we explore the basis of this defensive stream of consciousness pertaining to every word of the Bible, we find the two original admonitions, namely the one in Revelation 22:18: "I testify to everyone who hears the words of the prophecy of this book: if anyone adds to them, God will add to him the plagues which are written in this book" (written circa 96 AD) and the other one that is frequently noted from Deuteronomy 4:2 "You shall not add to the word which I am commanding you, nor take away from it, that you may keep the commandments of the Lord your God which I command you," you begin to see that these are warnings meant to apply only to the specific books therein. One must look at things in contextual application for them to make sense. The passage from

Deuteronomy was written at least a thousand years before Revelation, so obviously it is specifically referring to the Ten Commandments, not to someone adding to the text in general, otherwise Revelation and most of the Bible is disavowed. Furthermore, many biblical scholars have noted that 1 John was most likely written circa 95-110 AD, after Revelation; therefore the Revelation passage is likely referring exclusively to "the prophecy of THIS book," that is, Revelation, not the entire Bible in general.

The Roman Catholic Council of Nicea in 325 AD removed several original apostolic books from the original canon, including the book of Thomas (Didymos Judas Thomas), which has multiple similarities to the synoptic gospels of Matthew, Mark and Luke. Surely the Nicean council is culpable of taking away from the word of God. The Apostle Thomas's book was found near Nag Hammadi, Egypt in 1945. Archeologists have dated it circa 60 AD to 140 AD, but since there is no mention of the destruction of the temple in Jerusalem in 70 AD it was likely written before that. Notwithstanding the Catholic changes, wouldn't it have been nice if they had left the original New Testament writings put together at or near the time of Jesus alone to allow us to figure it out for ourselves, instead of removing whole books and adding others? Like the overall influence of your brother in law example, over time the frog put in warm water that is being gradually heated doesn't know he's being boiled until it's too late. One wants to believe every word in the Bible is inspired by God, but then we have to say the Catholics are in charge of a significant portion of the New Testament where they made changes. It's a paradox of thought because if changes are not to be made that rule has been violated, but if you are allowed to make changes as was done many times after the death of Jesus, can we keep making changes today?

Incidentally, one thing to remember is that when this was written in 2 Timothy 3:16: "All scripture is inspired by God…" the New Testament did not exist yet. Christ and the apostles were always referring to the Old Testament.

You are not insulting God by trying to better understand His word. That is why this section's lean toward Christianity is contrasted with the following section on the ancient Hebrews whose descendants became Rabbinic Jews; they wrote the truly God inspired Old Testament, especially the Torah; subsequently God completes His plan in the New Testament. It's important to see both sides to see the whole picture. After all, there seems to be at least a duality in this universe and much more in between. We see duality in the sun and the moon, male and female, old and new testaments, good and evil. Nothing is one sided, and often two come together as one. There can be no up without a down. There can be no heaven without a hell, otherwise there would be no measure for discernment or judgement, and everything would be the same- and life would be very boring if we were all the same.

Sir Isaac Newton's 3rd law of motion states "for every action, there is an equal and opposite reaction." This applies to a large degree to the duality of God's formations, and can be seen in Isaiah 45:7: "I form light, I create darkness; I make well-being, I create woe; I, Adonai, do all these things." Darkness and woe are negative, opposite reactions created as a result of His positive, good deliberations. A cup of coffee contains something we make, but when empty we don't say we made that emptiness- the latter is "created" or now exists because of the former.

With this same principle comes the existential involvement of Satan: he exists as a natural opposite of God. We use our freewill to exercise moral judgement to do the right thing, and thus defeat Satan's influence.

The most pervasive and consequential change the Roman Catholic church made was in 364 AD, when the Ecumenical Council of Laodicea officially got rid of God's 8th commandment to honor the sabbath day. 3. In Genesis 2:2-3 God stopped after He created the earth in six days. "Shabbat" means to cease, desist. He "blessed the seventh day and sanctified it" because He ceased to work, instead observing what He had done. Here is where God gave the sabbath to mankind, way before Israel even existed.

Emperor Constantine of Rome had decreed on March 7th, 321 AD that "all should rest on the day of the Sun." The pagans were already worshipping Mithra, the sun god on Sunday, so in a political compromise to unite Christians with pagans Constantine gave Sunday a new name "The Lord's Day," using the fact that Jesus had risen on the first day as a rationale. The Catholics have always believed that their pope has the authority to replace God's ordinations with man-made decisions, and Constantine was regarded as a pope.

God re-established the sabbath in Exodus 20:8 when He tells Moses to "Remember the sabbath day to keep it holy." In fact, the sign of the Mosaic covenant is the keeping of the sabbath by those who trust in Him.

In Rabbinic Judaism that which is within the Talmud asks us to accept imperfection and uncertainty (the "Talmudic Process") and to see the process of debating laws and texts as just as important as the finished result. That is how Judaism continues in seeking to understand God. However, they are never going to change the Torah, instead they study it by using logic to find the legal reality within it.

That's what distinguishes man from the animals, an ability to do critical thinking. I didn't say people are better than animals, just that they have higher order thinking skills beyond animals. Animals simply follow their instincts, but mankind has a heart that is desperately wicked, and that is why we need God. Once we have a relationship with God our mind aligns with our heart and the two form a renewed spirit of righteousness.

"Thus says the Lord, let not a wise man boast of his wisdom, and let not the mighty man boast of his might, let not a rich man boast of his riches; but let him who boasts boast of this, that he understands and knows Me, that I am the Lord who exercises lovingkindness, justice, and righteousness on earth; for I delight in these things, declares the Lord." (Jeremiah 9:23-24)

1. Steven E. Jones, "A Brief Survey of Sir Isaac Newton's Views on Religion," in *Converging Paths to Truth*, ed. Michael D. Rhodes and J. Ward Moody (Provo, UT: Religious Studies Center, Brigham Young University; Salt Lake City: Deseret Book, Salt Lake City, 2011), 61–78.

2. Sir Isaac Newton. *An Historical Account of Two Notable Corruptions of Scripture*. 1754. World heritage Encyclopedia WHEBN0000714104

3. Council of Laodicea, 364 AD, Canon XXIX (29)

Part II

What God is Like

Chapter 4
Geology and History

When your mind is trying to decide if you believe something or not most folks want to see more than a one-sided opinion on a matter. Since we have strong emotions there is the usual emotional or subjective reaction first, versus the objective or more empirical leaning deductive reasoning- but what if there is an area in between, sort of a gray area instead of just black or white sides, a blending together of phenomenon and fact in this mostly secular world. We should want everything to make sense, but there are occurrences that are unexplainable, yet they are part of our existence; this is where faith is born. There could never be one subjective standard that would suit everyone because each of us is independent from the next person. Free will and fate vie for reality in a tug of war, and this is why existence is perfect in its imperfection.

Most folks have the experience called deja vu when you feel you've been somewhere before, or have already seen an action play out and are re-living it again. Visible light is less than one percent of the electromagnetic spectrum, yet we know there are radio waves, radiation and much more than that- we just can't see them all. Or how about Kirlian photography, where a person's missing limb still shows up as an outline in the photograph? The truth is somewhere in between these two sides of life, the known and the unknown, and may not be as rare or as hard to find as one may think. The inquisitive mind that wants fairness looks for truth in all things before taking action. This was the case in the country of the United States that was designed based on truths that are "self evident." Truths of freedom, the right to live, and the right to worship are obvious and have been part of humankind's epistemology and ethics for over three millennia.

If you are a Christian then chances are your church (if you attend one) focuses mainly on Jesus Christ, or even the writings of the Apostle Paul. This chapter illustrates some major highlights of the first, original book of the Bible, the Old Testament, so the Christian can see and connect it to the origin and basis of the New Testament, sort of the other side of the coin, and therefore make informed decisions about their religious beliefs. Most importantly it will help you to understand God from a biblical anthropological perspective.

The ancient group of Hebrews who wrote the Pentateuch texts et al, knew about God, proceeding through the centuries to reiterate stories taken from actual events into His Word. God was promoting mankind through selecting certain historical accounts and weaving them through His chosen people the Hebrew culture. It's not fiction, it's the merging of science and faith blending together in that gray area mentioned earlier. The living God tells us with purposeful direction the contents of his book called the Bible, especially and initially in the Torah, then the Tanakh, or the twenty four books of the Hebrew Aramaic Jewish canon. The B'Rit Hadashah, or New Testament comes much later. This blending of historical facts and religion with logical reason gives prediction that comes to pass because it is God's will to make what He says happen, seen evidently in the continuation of God's chronicle from the Old into the New Testaments. To hold the view that these two books are of separate things is like the analogy presented by the brilliant Rabbi Kirt Schneider, where he describes returning home by taking a plane with a stop over in an international airport, but then missing your connective flight. To not give credibility to the theophany of hundreds of prophetic passages in the Old Testament fulfilled in the New Testament fails to see the natural unfolding of God's plan for the world.

Let us look at some of the characteristics of God according to what he chose to highlight in His Word: these are the truths of freedom, life and the importance of our worship of Him.

Of all the larger than life stories the tale of Noah (Genesis 5:29) and the flood is most widely known. This is because this story appeared in several countries between 2700 BC to 2000 BC, the oldest writing coming from a Sumerian account in 2300 BC, followed by a Hindu story called Gilgamesh whose hero was named Ziusudra written around 2150 BC, although the actual event is known to be much older. The Hurrians from the city of Haran in Babylon held the story before it was composed by the Hebrews circa 2000 BC, who assigned to it their hero named Noah. In all the archeological accounts this flood occurred at the Black Sea. Other cultures including ancient Greece had their stories of the flood as well.

According to geologists there have been at least two major floods in the last five million years. The one we are concerned about happened after the ice age circa

11,700 BC where a sudden loosening of sediments caused a major flood of salt water into the Black Sea between 7000 BC to 5500 BC near Turkey. At that time much of the earth's water mass was ice and the Black Sea was 400 feet lower than today. This had allowed the land to be used for farming, and the water there was fresh water that was running down from glaciers. The isthmus between the Mediterranean and Black Seas overflowed, and at a critical point a massive flood of sea water smashed into the Black Sea. This is scientifically accurate and has been proven by geological sediment layers. As ice continued to melt the entire known world eventually experienced floods. Therefore, although Noah may be thought to be an archetypal mythological character among other persons named by diverse cultural accounts, the flood itself was real. As an archetype Noah represents the sage who believed God when everyone else thought he was crazy to build the ark. The evident truth that God is illustrating is the right to life; however as the creator God supplies that right to those who follow His commandments and are therefore part of evidential truth. Your right to life is tempered by your free will. Noah chose to believe God and tried to warn the people that the flood was coming, but due to their disbelief they ridiculed and scorned him. The moral is if you trust in God you will live, but ignoring God for a prolonged period of time until your life digresses into continual sin causes the forfeiture of your life: imbedded into this story is the sinfulness that was widespread among the people, and the fundamental reason God destroyed them in the first place. So the Hebrews wove the moral aspect in as the causation of the flood. Noah and his family survived and he lived over 900 years. There are several countries who claim to house the tomb of Noah, including Turkey,

Iraq, Jordan and Lebanon, so he was likely a real prophet within the Hebrew culture.

No attempt at a character study of God would be replete without some rendition of His friend Abraham, the first patriarch of the Jewish people. Born in Ur a city of Babylon (today's Iraq), some historians say Ur was where Kuwait is now. Various accounts date his life between 1813 BC to 1200 BC; however it is most likely the earliest date (1813 BC) or even before that is accurate, because at the time of Abraham's young life Nimrod (in Hebrew means "the rebel") was the king, and he was the great grandson of Noah, born from the lineage of Noah's son Ham. Nimrod was the one who built the tower of Babel, which archeologists carbon date between 3500 BC to 3100 BC. According to anthropologists baked brick, which was used to build the Tower of Babel was not widely used until 3100 BC. There are accounts Abraham frequented the house of Noah as well, which places his lifetime closer to Noah's. Further corroboration that Abraham lived at this approximate timeline is that he was blessed by the priest-king of Salem Melchizadek (Genesis 14:18). It is plausible "Jerusalem" may have been derived from "Salem." More evidence of the timeline is in an ancient Ethiopian account about the Melchizedekites where Melchizedek worked with Shem, the blessed son of Noah to place the body of Adam into the Ark.

Melchizedek means "King of Righteousness," from the noun melek (king) and the verb sadeq (to be just). Exploration into the mystery of who Melchizedek truly was can be found by careful contemplation of the story in Genesis 14:37, when Abram, upon finding out that his nephew Lot had been taken captive by Omer the king of Mesopotamia, mustered together his own force of 318 men to go save him. Through a guerrilla warfare tactic of dividing his men into groups, surrounding the tail end of Omer's troops and continually attacking at night Abram was able to secure enough of a victory to rescue Lot and others. Shortly afterwards, among the kings that congratulated Abram appeared the king of Salem Melchizedek with bread and wine, a priest of El Elyon (God most high). Melchizedek blesses Abram, who is actually a descendant of Melchizedek through the line of Seth, Noah and Shem. Among ancient sages and scholars of Torah, Melchizedek is actually Shem, which in Hebrew means "name." Shem was the middle son of Noah, and Noah blessed him: "Blessed be the Lord, The God of Shem" (Gen 9:26). A name reflects character, and Shem was blessed as the one the promised line of the Messiah would come from. Shem outlived Abram, who God renamed Abraham.

The story of Abraham that is the catalyst which first causes God to choose him is the one about how an adolescent Abram destroyed his father's idols. His father Terach made idols for a living, but Abram knew about God and one day when his father was out he smashed all the stone idols save for one, the largest idol. Abram put the hammer in that idol's hand, and when his father returned explained that the idols had a fight and the large one killed the others. Terach replied angrily that was not possible because these stone idols can't do anything, thus proving Abram's point that the idols were dead stone figures and not living Gods. Many years later God changed Abram's name to Abraham. Although this story of the young Abram is not in the Old Testament it is in a Midrash from the Talmud. 1.

The most important event in Abraham's life occurs when he is about to sacrifice his son Isaac, but God stops him and tells him to sacrifice a ram that was caught in the thicket instead (Genesis 22:1-14). This is a prophetic foreshadowing of God sacrificing His own son Yeshua (Jesus) which occurs 1800 years later, with the exception of course that Jesus was actually killed. Abraham went on to live 175 years and is buried in Hebron, a city in the southern West Bank in the Judaean Mountains, 19 miles south of Jerusalem.

1. *Sefaria.org/sheets/24117. Genesis Rabbah 38.13, 300-500 CE*

Chapter 5
Moses and the Rock

Water can be deadly but it is also the most important element for life to survive. A person can go without food for three days but cannot go without water for more than a day. We have seen that God's voice is like the sound of water, and this quality is also ascribed to Jesus in the context of Revelation 14:2: "And I heard a voice from heaven like the sound of many waters..."

It is so important that we credit the power of living water to God's glory. The fact that it wasn't done at the crag of Meribah by Moses is attributed to be a direct cause of his death and of his brother's as well, the high priest Aaron (Numbers 20:1-12). After God told Moses to speak the command to the rock to gush forth water, instead he disobeyed God by speaking to the descendants of Jacob and striking his rod on the rock twice. Although this resulted in water gushing out of the rock anyway Moses failed to give God the glory, and this resulted in God becoming very angry. Moses and Aaron failed to show the proper confidence in God in calling water from the rock by not following directions. The wrath of God is terrifying, and you would be smart to avoid His anger: "The fear of the Lord is the beginning of wisdom, and the knowledge of the Holy one is understanding" (Proverbs 9:10) In other words you better be aware of who is in charge of your life or you could lose it. Although God's punishment was severe, throughout the time Moses was alive he was essentially God's most trusted bond- servant and He loved him. Moses the law giver was extremely important because he was the first to establish a covenant with Yhwh that produced religious law, the foundation of a legal system followed by the prophets and priests, essentially based on the Ten Commandments.

When God planned Moses' death He told him to prepare Joshua to take his place. Although Moses really wanted to enter the promised land that was taken away because Moses, in a momentary lapse of faith withheld glory from God. So then you ask if He loved him then why such a harsh sentence- why end his life? An alternative explanation that some biblical scholars have positioned is that since Moses was one hundred twenty when he died and all the first generation of escapees from Egyptian captivity had already died, it was only befitting that Moses should die with them. Thus, Moses' death may not have been only a direct result of his disobedience at Meribah-kadesh, but also a natural generational appointment.

From an empirical, historical perspective, the story of Moses in Exodus, written between 800 BC to 600 BC is deemed by some to be a cultural myth that has been handed down through the centuries. Indeed it existed in Persia for two thousand years in the historical account of Sargon the First before it was adopted into the Hebrew culture. The exodus actually did occur though, between 1312 BC- 1250 BC, albeit it was probably on a much smaller scale than told. The mysterious conundrum about the Moses exodus story is the lapse of time between the event and the writing of the Exodus book in the OT. It is reasonable that he lived in ancient Egypt and led the exodus, then his name was assigned by the ancient Hebrews as the writer of the Torah several centuries later. In Exodus 12:40 we see the Hebrews were in Egypt for 430 years until the exodus.

There is Egyptian documentation that the famine occurred in 1225 BC and brought a group of people from the tribe of Jacob in the land of Canaan into Egypt seeking food. Joseph was instated into the house of Pharoah at this time. A cultural conflict revolved at first around the Hebrews eating of sheep which was considered a sacred animal by the Egyptians. This was one of the reasons the Hebrews were relegated to be day laborers, and when leprosy broke out among the working class an Egyptian priest named Osarseph rose in their defense. The problem with this timeline is it is backward to the sequence of events where first came the famine, then the exodus supposedly 430 years after that.

The name Moses is Egyptian: the Jewish historian Josephus (37 AD -100 AD) said in Egyptian "mo" means water, and "uses" means saved. Another translation in part emerges from the Coptic "ushe" which means rescued. In ancient Egypt there is a true story of a rebellious priest named Osarseph who wanted to worship one god, and so rejected the polytheism of Egypt. Osarseph's existence is recorded historically (however it's mostly lost) first by the Egyptian historian Manetho (circa 250 BC). Manetho's tale was recounted by Josephus; the story tells of a renegade Egyptian priest who leads an army against the Pharoah. Manetho told Osarseph's group was exiled out of Egypt and went to Judea where they then founded the city of Jerusalem. At the end of the story Osarseph changes his name to Moses. 1.

The Old Testament story of Moses written around 600 BC by ancient Hebrews connects to the timeline of Joshua 17:18, where Joshua saw that the Canaanites who lived in the valley could not be defeated because they had "iron chariots." This might indicate the time period within the late iron age, which was from about 500 BC to 332 BC. However the Canaanites are associated historically with the Bronze Age, so obviously a transition from bronze to iron is developing among several peoples such as the Hittites who lived in what is now Turkey and are credited as the first to make steel by heating iron with carbon.

Even though the Penteteuch was written several centuries after he died it is apparent that Moses did lay the groundwork of Biblical precepts by providing important documentation for the main writers of Genesis and the other Penteteuchal books to keep and elaborate on later. The mathematical timeline inconsistency appears again here, for how can Moses have worked with Joshua during this period when he already led the exodus 600 years earlier? Curiously, Hebrews 3:5 alludes to this time discrepancy: "Now Moses was faithful in all His house as a servant, for a testimony of those things which were to be spoken later." Albeit the true answer is the story of Moses and Joshua must have occurred together during the exodus circa 1290 BC; this would be before the time of what is considered the rise of Israel which is noted during the second part of the Iron Age (1000 BC-550 BC), not the third or late Iron age. Egypt had lost its domination in the region allowing for settlements to establish along the area known as the levant (French for "the rising") along the east Mediterranean coastline. The Hittite empire had already diminished as well (around 1180 BC), most likely due to the famine.

Regarding the exodus there is much accretion to the accounts of this particular event, coupled with many contradictory stories about Moses' career in general. The most consistent recurring Moses story reiterated by many historians tells us he was an infant in Egypt and brought up by the daughter of a king. The legend of the Babylonian king Sargon (3800 BC) whose mother had placed him in a basket into the river is cited as the oldest, original source of this part of the Moses story.

Moses died in the territory of Reuben and is buried four miles away in Gad on Mount Nebo in what is known today as the country of Jordan. He is in an unmarked location, but other than that belief there is no archeological evidence of his remains in Israel.

Why establish this story, and how has the story of Moses endured for centuries in so many different cultures? It's God who makes the self-evident truth of freedom manifest through Moses, the archetypal ultimate hero type who frees His people, and through God's will Moses delivers. First he frees them, then brings them toward a place where they will live in harmony with God. When hardships occur the people lose faith, complain and worse, they build an idol to a false God. Sadly, even Moses momentarily forgets his faith at the Meribah rock. This is because unlike Jesus the god-man, Moses is only a man. However, much was achieved by him by the time Joshua takes over, and clearly God shows favor and love toward Moses throughout the Penteteuch. Hastiness and loss of faith caused trouble, even Moses' death. On a personal level what God is saying here is without faith and obeyance in Him your soul will die, but if you have faith and obey God your soul will live. In The Song of Moses the namesake tells the Hebrew people "For it is not an idle word for you; indeed it is your life." (Deuteronomy 32:47)

The story of Moses has classical elements such as the reversal of Situation (similar to Reversal of Intention), and Recognition; these were the most important features stated by the philosopher Aristotle (385 BC-322 BC) 2. of Greece for a story to be significant. Since the OT Moses stories were written circa 600 BC they would have been just a few centuries before Sophocles (circa 496 BC- 406 BC) who wrote a play called Oedipus the King, credited among the classical Greeks as the best story ever written. Oedipus runs exactly into his fate while trying to run away from it (Reversal of Intention) and is recognized later in the story by his birthmark (Recognition). The timeline is close enough that the information could have been easily transferred from one culture to another and borrowed by the Greeks. As a baby Moses is sent down the river in a basket but is intercepted by Pharoah's daughter who takes him and raises him in the palace. During his upbringing he realized he was a Hebrew (Recognition).

Furthermore, the third element Aristotle said made for a most excellent story is the Tragic Incident. In the case of Oedipus he marries his own mother, corresponding to the ancient cult action of the Babylonian king Nimrod, but not so with Moses. The major dynamics in Nimrod and Moses coalesced into the Oedipus story.

In the Song of Moses in Exodus where Moses addresses the assembly of Israel he tells the story of how God found and chose them, protected and loved them, but they turned away, forgetting God. God complains that these people have no understanding or gratitude, but He continues to watch over them because He's committed and does not want their enemies to destroy them. It is notable that the word "rock" as a description of God appears here eight times. Obviously this emphasis helps explain why it was so important that Moses speak to the rock at Meribah-kadesh. The discontent expressed by God in the Song of Moses is nevertheless followed by a blessing given by Moses to the twelve tribes of Israel. (Exodus 33)

Avoiding the wrath of God by fearing and respecting Him is a message throughout the Bible, and particularly through the catalyst of the Song of Moses, which appears again in Revelation 15:3-4. Here in Revelation 15:4 the Song of Moses briefly expresses the theme of fear in Proverbs 9:10 when it says "Who will not fear, O Lord, and glorify Thy name?" In a reflection of our own lives if you don't fear consequences how will you keep any sense of right and wrong to govern your actions? Without law people have no moral compass to keep them from killing and doing crimes in general.

Part of God's Old Testament personality is he gets angry and jealous if someone takes or gives the credit, otherwise known as glory to someone else other than Him. "You shall not worship them or serve them; for I, the Lord your God, am a jealous God…" (Exodus 20:5) Revenge also belongs to God who is omniscient, patient and knows He will triumph: in the Song of Moses He warns "Vengeance is mine, and retribution, in due time their foot will slip…" (Deuteronomy 32:35) So you don't get away with anything- God is like karma: what goes around comes around. God cannot compromise His holy judgement, and everything must be weighed in the balance (Proverbs 16:18). Furthermore, He has the luxury of being "slow to anger" (Psalm 103:8) for our time is going much faster than heaven's time: by the time an hour goes by in heaven 41.6 years has transpired in earth time. God can wait us all out, and this may show why He is so patient.

There is a New Testament scene where Jesus asks His disciples who do people say he is? They give him various answers, so he asks them again "But who do you say I am?" Peter answers this time saying "Thou art the Messiah, the Son of the living God." Jesus blesses Peter, telling him that the way Peter understood this was revealed to him by God Himself. Jesus then adds that Peter is a large rock, and upon this bedrock "I will build my church; and the gates of Hades shall not overpower it." (Matthew 16: 13-18) The point Jesus is making here has to do with the transformative power of God to enlighten the saved, just as God had enlightened Peter. The incident itself is not so exceptional as to say Jesus is appointing Peter as the head of His church, considering there have been many other supernatural and remarkable events too, such as the revelations given to John, and the vision of the risen Messiah seen by Mary Magdalene at the tomb. Peter is an apostle equal to John and Mary and part of His inner circle; Jesus is the nexus for the divine blessings of His apostles. A large rock is not the same thing as a bedrock, the former a single entity while the latter is an entire foundation. The use of the word "bedrock" alludes to a personification of God, and Peter is one important rock that sits upon it. A corollary to this story is found in John 1:42 where Jesus tells Simon Peter "you shall be called Cephas" which in Aramaic means "stone." Once again Peter is a singular stone or rock, important but not the entire foundation of Jesus' church.

An incidence of God being noticeably absent from a main prophetic fulfillment scene occurs at the final minute of Christ's death when he calls out "My God, my God, why have you forsaken me?" (Matthew 27:46) It could be that God is too holy to allow Himself to be pulled into death. It wasn't part of the plan that God would come to the rescue on the cross. Remember Jesus explains to His disciples the Holy Spirit or Advocate is in heaven and he has to ascend to get him. After Jesus' life was ended He triumphed over death, as the mission proved successful. "The Kingdom of the Father is like a certain man who wanted to kill a powerful man. In his own house he drew his sword and stuck it into the wall in order to find out whether his hand could carry through. Then he slew the powerful man." (Thomas 98). God already knew His son could do the job but this time He exercised the full power of retrieval of Messiah's soul from the low point, so there would be no question of His power. "It is finished" said the Messiah. (John 19:30)

1. Osarseph. World heritage Encyclopedia. WHEBN0002030540. © 2021 World Library Foundation

2. Aristotle. Poetics (384 BC-322 BC)

Chapter 6
Judaism and Christianity

Many people hold the opinion that the Old Testament God is so different from the New Testament God, but not really. Jesus reinforces the Old Testament, or "Tanakh" (Hebrew) when He comments that He came to fulfill the law, not abolish it (Matthew 5:17). Jesus said He came to save many, but not everyone. When God is giving the Ten Commandments He says He is "showing lovingkindness to thousands, to those who love Me and keep My commandments." (Exodus 20:6) So that's not everyone only those who follow God's will.

"To elaborate on a point mentioned earlier (in chapter 2), some Christians may mistakenly think they only need to pay attention to the New Testament and not the Old because they believe Christ replaced the law with grace, and that's all you need for salvation. However, in Matthew 5:17 Jesus proclaims "Don't think that I have come to abolish the Torah or the Prophets. I have come not to abolish but to complete." The Greek word used here for complete is "pleroo," which means to fill with meaning, to make abundant, and bring to the fullest extent. What it doesn't mean is to replace.

Furthermore, God never says we are to choose grace over law. This is a false dichotomy. In fact, grace first appears in the Old Testament in Genesis 6:8: "But Noah found grace in the sight of Adonai." So it has been grace all along, as is always the case when we have faith in God: we are then graced by Him with righteousness. Grace is a main tenet of Judeo-Christianity in both testaments.

In Exodus 23:18 and Leviticus 2:11 God orders that anything served to Him is not to contain leaven, also citing leaven as a way to have the Israelites remember their hasty exit from Egypt, in that there was no time to use yeast to make the bread rise.

The subject of leaven has a role in prophetic prophecy regarding Jesus, as He called himself "the bread of life" (John 6:35). Leaven adds air to the dough, puffing it up so it appears greater in content than it actually is: this is why leaven represents false pride, hypocrisy and sin of the flesh. Furthermore, leaven symbolizes corruption and contamination, while unleavened bread represents purity of mind and heart. This is why Jesus warned to beware of the leaven of the Pharisees and the Sadducees in Matt. 16:5-12, Mark 8:15, & Luke 12:1. In the Luke passage Jesus specifically equates leaven to their hypocrisy.

In John 6: 51 Jesus explains further: "...and the bread also which I shall give for the life of the world is my flesh." Jesus was sinless, a personification of unleavened bread, containing no corruption. This lends credence to why when He died he did not decay and came back to life. Jesus was killed as the Passover lamb. He died on a Thursday, the eve of Passover 1. and was buried in a tomb that evening at the beginning of the Feast of Unleavened Bread. He rose early Sunday on the day of the Festival of the First Fruits. His death and resurrection are the atonement for sins.

Yom Kippur, a very high holiday appointed by God in Leviticus 23, is about atonement. In Yom (day) Kippur (covering) God requires that we cover our sins by atoning for them. When you engage in connecting Yom Kippur to Yeshua Hamashiach (in Greek Jesus the Christ) you will see another dimension to this high holiday, and deepen your understanding of the Messiah.

Christians tend to hold the one view that faith in Jesus is all you need to get saved, in that when He was killed He took the place of the Passover sacrifice once and for all, replacing God's covenant requirement for the blood sacrifice of the Old Testament. Since people were wicked and sinful, nothing could be done in the old days before Jesus came to appease God's anger except the death of an innocent lamb as a temporary measure. Things have changed: now is the time of the new covenant in the New Testament, or "B'Rit Hadashah" (Hebrew) whereas according to John the Baptist Jesus took care of the sacrifice permanently (John 1:29); that is also why He is called the lamb of God in the Song of the Lamb (Revelation 15:3).

Christianity says God has one son, but Judaism says although He has countless sons and daughters the Israelis are His firstborn, while Christian gentiles are also His children. God wanted to help mankind and he had to go through someone, so he chose a wise and gentle people to represent his plan. If He hadn't done this it may be that we would all be lost. It makes sense that Jesus Yeshua was born within the Jewish people as Israel is the place God has invested his plan, and heaven is where through faith people from both parties are saved into the next life as the fulfillment of his plan.

God condemns sin, but He also hates death and would prefer all souls, even sinners to live: "I take no pleasure in the death of the wicked, but rather that the wicked turn from his way and live." (Ezekiel 33:11) It's pretty clear He despised death to the point where He did something about it, announcing early on through His prophets that a savior was coming through the line of King David: "In those days and at that time I will cause a righteous Branch of David to spring forth; and He shall execute justice and righteousness on the earth." (Jeremiah 33:15) Then, again in Ezekiel 34:23: "Then I will set over them one shepherd, my servant David, and he will feed them himself, and be their shepherd." God is telling the world that He's providing a way to save our souls through the southern kingdom, the lineage of Judah. In fact the word "Jewish" etymologically comes from the name "Judah." Finally, Jesus fulfills his destiny, naming himself as the Messiah: "I, Jesus, have sent My angel to testify to you these things for the churches. I am the root and the descendant of David, the bright morning star." (Revelation 22:16) The angel He's referring to is the Holy Spirit. In Hebrew Jesus is "Yehoshu'a" which is Joshua, and "Christ" is a Greek word that means "Messiah."

Just as the warning of God's wrath to non-worshipers is followed by a blessing of God's love by Moses, God balances His own emotions by telling us "But the lovingkindness of the Lord is from everlasting to everlasting on those who fear Him" (Psalm 103:17) He doesn't quite apologize when He announces "In an outburst of anger I hid my face from you for a moment, but with everlasting lovingkindness I will have compassion on you." (Isaiah 54:8) As one can see God has intense sensitivity regarding emotions of anger and jealousy. If you have a cherished friend you are not going to provoke him by making him angry or jealous, unless you are trying to get rid of him. No, we all need companionship, as we were not made to be alone. This is why God emphasizes the importance of Judeo-Christian fellowship; that is to say when you gather together in the name of God you are affected by His Holy Spirit, and therein God bestows His blessings of grace on the congregation.

One example that illustrates the personality of God is found in the story of the prophet Jonah (786-746 BC), who refused to listen to God's command to travel to the Assyrian-Babylonian city of Ninevah (today it is Iraq) to prophesize their imminent doom due to their wickedness. While trying to flee God's command Jonah was swallowed by a large fish (Jonah 1:17) and was vomited out three days later. When God gives you a directive you need to do it or suffer the consequences of his wrath. Jonah found this out the hard way.

There are two occurrences in the Jonah story that parallel Jesus' life: Jonah slept during a violent storm on the boat in the Mediterranean Sea like Jesus did in the Sea of Galilee boat (Matthew 8:24), and Jonah was resurrected three days after he died like Jesus was. The differences are that Jonah was thrown overboard and the storm quieted, while Jesus himself calmed the storm. Jonah survived death from the fish while Jesus came back to life from a tomb- both after three days. These are significant events because Jonah's story is a foreshadowing of the Jesus story 750 years before Christ. The omnipotent personality of God is shown in the moral message and evidential truth of the prophet Jonah story: obey Him or forfeit your life. After Jonah went to Ninevah and warned them the people there repented from their sinful ways and the city was saved from destruction.

Jonah really existed, as his tomb was found in Iraq above a 2700 BC palace where a known king ruled about a hundred years before him. Incidentally, in 2014 the militant Islamic group ISIS destroyed the tomb of Jonah.

Have you ever known someone who brings out the worst in you; when you are around such a person you seem to be angry a lot, and you realize this is not the real you, that your naturally good side is somehow suppressed because of the negative emotions constantly at play - this was God's dilemma: He wanted to show love, joy and mercy toward his people, but they didn't remember his kindness and kept on sinning. Nevertheless, because he chose them His love prevailed, never quite destroying all of them, as a remnant always endured.

All four stories of the prophets Noah, Abraham, Moses, and Jonah are in the Babylonian Talmud, while the first three are in the Torah, the first five books of the Bible (also called the Pentateuch). While the Hebrew OT Bible consecrates these stories the intention was not to give precise, accurate history. God gives us free will and enough of a guideline on how to live our lives, and part of the plan is to see who becomes faithful and obedient towards Him.

Free will in man allows choice of at least two sides, a good one and a not so good one for example. Regarding the tree of the knowledge of good and evil we know that Eve replied to the serpent in Genesis 3:3 that God had told "You shall not eat from it or touch it, lest you die." However, God never said not to touch it; He told Adam only to not eat of it, before Eve was created, so it was

Adam's responsibility to inform her. Eve lies when she says they can't touch it, before she even eats of the tree; or is she only repeating what Adam told her? This shows that God put an inclination, or natural tendency in humans to lie, or at least embellish- we seem to have a natural tendency to tell stories. One can't help but wonder if this life is a test to overcome our fallen nature.

The earliest record of the term "Israelites" is from an Egyptian tablet dated circa 1200 BC- these were the people who were formally Canaanites who left urban areas to become self-sustaining agrarians. By the time Joshua arrived into the promised land from the exodus out of Egypt (historically noted to have been circa 1290 BC) he was told by God to defeat the Canaanites for the sake of the Hebrews to inherit the land; thus it appears the two groups merged into one. It is likely the Hebrews had returned to reclaim the land that had been theirs before the Egyptian habitation. One synthesis between the two groups that did occur is the Hebrews influenced the Canaanites to convert from idol worship to monotheism "Yahwism."

The Babylonians under king Nebuchadnezzar II destroyed the Hebrew temple on the Temple Mount in 586 BC and captured many as slaves- there is archeological evidence of this destruction. The Hebrews returned to Israel about 50 years later. Concurrent with this calamity the Hebrew people adopted the stance that God wanted them to get rid of their female fertility Goddess Asherah (who supposedly had been God's companion) and worship only Him. There are thousands of figurines of Asherah that have been found carbon dated between 500 BC to 900 BC, and inscriptions of her date into the 8th century BC. This deliberate religious change toward only a single God is reflected in the first two commandments that direct monotheistic worship. However biblically Moses received the Ten Commandments after the exodus which was circa 1290 BC, so the time period does not match perfectly the deletion of Asherah, but does match the Rabbinic tradition of the beginnings of the writing of the Torah.

After King David ruled in 840 BC and his son Solomon, the Hebrew kingdom suffered a civil war where it was split between north (Israel) and south (Judah). Subsequently came King Josiah of Judah, (reigning from 640 BC to 609 BC) who permanently and completely did away with the worship of Asherah.

Last but not least there is the oldest story in the Bible, the book of the prophet Job, not found in the Pentateuch but in the Talmud. The premise is the question all believers ask, and that is why would a just God allow tremendous suffering in good people's lives. This is the single most asked question among all people for all time and even today, whether they are believers or atheists and this is why the book of Job is very important. When a person denies God's existence they often ask "if God's in control then why does he allow evil? He can't be very good because of all the turmoil in the world." This is a fundamental misunderstanding of God, for He never promotes evil, He allows free will. It is mankind who does evil things. Remember we are not in heaven yet- we are still in this place that is a mixture of good and bad. We're not here very long either- only long enough to figure out and choose virtue over sin, and to find virtue we need God.

There are several arguments and reasoning platforms that transpire in Job's symposium, but the reader must remember that everything happens under the umbrella of Satan interfering in Job's life; this evil one is the causation of the suffering and great tribulation, not Job. The evident truth is the right to pursue happiness but not the guarantee of it. In conclusion, we are to realize that here in our mortal lives evil exists, and even if you are upright before God you'll still have bad stuff happen to you. Job is faithful to the end, as we must also be. God makes good and restores most of what Job lost. A New Testament comment on this is in James 5:11: "You have heard of the steadfastness of Job and have seen the end of the Lord that the Lord is full of compassion and is merciful."

Rabbinic tradition is that Job actually did exist, and the book was written before the sixth century BC, although the story itself is older. Is it not known if it was written by Job or Moses, but Job's wife was one of Jacob's daughters, so the timeline of his existence was between 1836 BC to 1689 BC. Jacob of course was the grandson of Abraham. Job was born in Uz, thought to be today in southwestern Jordan near the border of Israel. He may be buried on a hilltop 16.76 miles from Salalah, in the Arab country of Oman, approximately 1500 miles from Uz.

God seems to like irony. Perhaps it's to show he is always in charge, and His will comes from Himself, not the perversions of mankind. The evil Haman was hung on the very gallows he had built to hang his enemy on- this was so unexpected (Esther 7:10). When King Saul's prophet Samuel goes to Bethlehem to find which of Jesse's sons God wants him to anoint as the next king of Israel God has Samuel choose the smallest, weakest looking one, the shepherd boy David (1 Samuel 16:13) The devil thinks he's won against God's son Messiah Jesus when He is crucified, but the very opposite is happening because Jesus defeats death and comes back to life and now the devil has lost his grip on the souls of this world. There are countless examples of irony in both Testaments. God may be amusing himself through irony when the unexpected comes true.

Another example of God's humor is in 1 Samuel 5:1-5, when the Philistines capture the Ark of the Covenant from the Israelites and place it in their temple to honor their idol Dagon. The next morning they find Dagon on his face, then the following day find Dagon face down with his arms and head cut off. God thought it was amusing to put the false idol Dagon in a submissive position before His Ark. Why should God be in distress over the affairs of man-rather "He who sits in the heavens laughs, the Lord scoffs at them." (Psalm 2:4)

In the Talmud we can clearly see the values of free speech and persuasion: the whole "Talmudic Process" of written tabulations of the Mishneh, and the notations of dialogue in the Gemara stand as the very essence of that. From the time and even before the Maccabees (a ruling family of the ancient Hebrews) logical reasoning and debate was used to find a legal base of the Torah; this to evolve an understanding of how God operates. God likely has a sense of humor while advocating a give and take process to understand Him; otherwise there would be little point of giving us free will and speech. He does not have a heavy hand like the God described by the 18th century preacher Jonathan Edwards in his speech "Sinners in the Hands of an Angry God" (1741), full of only anger towards man, although He can get to that point. God's real nature is loving and full of joy.

This chapter is not intended to represent a complete study of Bible origins. It is presented to highlight some of God's main ideas of how he developed His relationship with mankind. This book is an exegesis based on the earliest available anthropological components, including history, geology, science and religion.

1. *Babylonian Talmud: Tractate Sanhedrin, folio 43a, 33-34*

Part III

For the Love of Jesus

Chapter 7
Mary Magdalene

There is a mystery about Jesus in the question of where he spent his teen age years into his twenties. A plausible location is Capernaum, a city far north along the northwestern shore of the Sea of Galilee near the Jordan river. It is about eighty-five miles from Jerusalem and thirty miles from Nazareth, along the trade route of the time. It was the home of Jesus's boyhood friends and his chosen first disciples, the brothers Andrew and Simon Peter who were fishermen. It is likely Jesus practiced preaching in the synagogue that was conveniently located eighty-four feet from Peter and Andrew's house where he often stayed.

This ancient trade route also passes through Magdala where Mary Magdalene was from. Mary Magdalene was one of the disciples of Christ, likely a friend of his mother Mary's, and did produce a book The Gospel of Mary in the early church. However, it is assumed her book was removed even before the first council of Nicaea in 325, concurrent with a false instatement consolidating her existence with a prostitute also named Mary. These were two different women, but again, early church hierarchy, not the least of which was Pope Gregory in 591 A.D., decided it was easier to omit her writings by defaming her. Furthermore, she is neither the woman who washes Jesus's feet in Luke 7:36, nor the mentally ill possessed woman in Luke 13:10-13. Women were becoming too powerful in the early church; this in the eyes of male patriarchy of the Roman Catholic church, who were competing for followers with the Gnostics, the followers of Mary Magdalene.

It is likely that Mary Magdalene was ten years or more older than Jesus and was his benefactor. Her book was discovered in 1896 in upper Egypt, and what remains of it can be found in the Berlin Gnostic Codex. 1. Mary Magdalene was the first apostle to see the risen Jesus at his tomb and she heard his instruction to go tell the other apostles he was alive. At the tomb she asked him how does the visionary see the vision, with the soul, or with the spirit? She was asking how was it possible that she was seeing him. Jesus answered her "It is not with the soul that one sees, nor yet through the spirit, but by the mind between the two. This sees the vision, and it is that by which I am speaking with you now, O Mary."

(Mary Magdalene 10:6-7) One explanation is that each of us has three parts, the control center of origins being the mind. In the mind is where choices begin that lead to actions, beliefs and memory. Those experiences in life manifest into strong feelings of the heart, such as love, or even resentment. Altogether they coalesce into the forming of the spirit. Jesus is presenting metacognition in that our mind is aware of what we do, and that becomes what we habitually do, and so our soul is formed thusly. "For as he thinks within himself, so he is." (Proverbs 23:7)

As the birth of Jesus was as humble as could be, so conversely his violent death showed the wicked and cruel depravity man is capable of. These extremes were applied so that evidence was provided for God's judgement, which is then completely fair against sin. Evidently Jesus died on April 3, 33 A.D., according to the prophet Daniel's prediction 500 years earlier. Speaking of Daniel, in 11:31 we read of "the abomination that causes horror" coupled with Matthew 24:15, which tells us this abomination will be "standing in the holy place" as a sign before the end of the world is to come. We know that the holy place is Mount Moriah, and the rest is a mystery that the reader, with a little study can uncover and figure out.

Of the three synoptic gospels Mark and Matthew were the first to be written, circa 40-60 A.D., before the destruction of Jerusalem by the Romans in 70 A.D. According to Mark and Luke Jesus lasted on the cross for six hours on that Thursday morning, from nine AM until three PM. Darkness overcame the land from noon until three PM, at which the latter time he cried out "My God, my God why have you forsaken me?" (Mark 15:34) It may help to understand that in ancient Israel the days were metered out from sunset to sunset. The evening of Thursday begins Friday, then subsequently through day three: Sunday "the first day of the week..." (Luke 24:1, Mark 28:1, Matthew 28:1, John 20:1) when Jesus Christ rose from his death alive again. There are several eyewitness accounts from numerous people who saw, heard and touched Jesus's resurrected body for forty days after he died. On that same first Sunday of his resurrection Jesus appeared to Mary Magdalene, a group of women, Simon Peter, to people on the road to Emmaus, and eleven disciples. On the eighth day after Jesus had risen the disciples called a meeting in their regular meeting place, the upper room and locked the door for fear of Roman persecution. As they were sitting at the table talking Christ appeared in the room saying "Peace be to you." (Luke 24:36) At first the disciples were very scared thinking they were looking at a ghost, but Jesus assured them it was he, having them touch his wounds, and even eating a piece of fish. Finally, Christ ascended to heaven on the fortieth day at the Mount of Olives Cemetery.

1. *The Gospel According to Mary Magdalene. The Gnostic society Library. Berlin Gnostic Codex (Papyrus Berolinensis 8502)*

Chapter 8
The Virgin Mary

Parthenogenesis, or a female's ability to self-conceive is already taking place among several species of reptiles and insects. 1. Science today has not realized the full potential of life forms that can do this, but theoretically it could account for part of the explanation of what happened two thousand years ago in Israel to Mary, the mother of Jesus. A scientific fact and physiological anomaly is that about one in every five thousand middle eastern women are hermaphrodites. That is, these people are born with both ovaries and testicles, and can self fertilize; however they are also born without a uterus. There is only so much room in the pelvis and a uterus would take up a significant area. Therefore the zygote, when having no base to attach within an uterus simply washes away, never developing into an embryo.

In Mary's case the odds were probably greater than one in a million of the one in five thousand delineation that are born with unique and necessary anatomical abnormalities that included a uterus making it possible to have a baby. It simply was never going to be likely for there was no precedent occurrence. Incidentally she did not have the male member part either as some hermaphrodites. When one ponders the math you see that the odds are so great that in all probability this mutation would not occur again, something like one in five million hermaphrodites. It was so rare that it could give credence to the belief that it was a supernatural will of God that it happened at all. There are no hermaphrodites on record today that have had the same specific set of sexual reproductive organs that Mary must have had. She was only a teenager, and this is significant because she had only recently reached puberty whenas her sex hormones became active. God exercised his will in a timely fashion in this very rare immaculate conception. A paradigm shift here brings us the concept that science and religion are not mutually exclusive- they are one.

When the time was right God intervened by sending a messenger angel to Mary who told her "The Holy Spirit will come upon you, and the power of the most high will overshadow you..." (Luke 1:35) The immaculate conception was a multi-dimensional event.

The birth of Jesus was foretold hundreds of years before it occurred. An amazing aspect of the existence of Jesus, who was to become the Christ was prophesized by Isaiah seven hundred years before his birth: "Behold a virgin will be with child and bear a son." (Isaiah 7:14) This same prophet also said Jesus would descend out of the house of David "then a shoot will spring from the stem of Jesse, and a branch from his roots will bear fruit." (Isaiah 11:1) Jesse was of course the father of King David.

The prophet Micah stated Jesus would be born in Bethlehem Ephrathah: "But as for you, Bethlehem Ephrathah, one will go forth for me to be ruler in Israel." (Micah 5:2) Bethlehem Ephrathah is the very old name for this area in Judah directly south of Jerusalem, also the birth place of King David.

Mary's cousin was Elizabeth, the mother of John the Baptist, so Jesus was likely a second cousin of John the Baptist. Elizabeth was a descendent of Aaron the brother of Moses and was married to the priest Zechariah. Mary and her husband Joseph were both from the tribe of Judah: she was a descendant from David and Bathsheba's third son Nathan, while Joseph came from the line of Solomon, King David's fourth son with his wife Bathsheba.

Zechariah shows us God's favor toward Israel in the Daughter of Zion prophecy that Mary will conceive the savior "Sing for joy and be glad, O daughter of Zion; for behold I am coming and I will dwell in your midst, declares the Lord." (Zechariah 2:10) Then "And the Lord will inherit Judah as his portion in the holy land, and will again choose Jerusalem." (Zechariah 2:12)

After Mary was visited by the angel Gabriel who told her she was to give birth to God's son she went to Elizabeth's house to tell her cousin the blessed news. The baby already inside Elizabeth turned inside her womb, and feeling this Elizabeth said to Mary "Blessed are you among women, and blessed is the child you will bear." (Luke 1:30-42) Elizabeth herself was six months pregnant with John the Baptist, born about six months before Jesus. Mary stayed with her three months until Elizabeth gave birth. At this time Joseph had been planning to disavow his engagement to Mary because he thought she had been unfaithful to him. What brought Joseph back to Mary whereas he married her was a dream he had where an angel told him she was pregnant from God, explaining the child was to be the fulfillment of the prophecies promised by God. After the birth of Jesus Joseph and Mary had several children, half-brothers and sisters of Jesus, including James, Joseph, Miriam, Ruth,

Martha, Simon and Jude. Incidentally, Joseph was over twice the age of Mary, likely in his late thirties.

People sometimes complain about the fact that Jesus being born into poverty was not commensurate with a demi-god savior, and this is called the mystery of the incarnation. You see, it was necessary for God to appear into a lower social status, for God's heart is with the poor. "Has not God chosen those who are poor in the eyes of the world to be rich in faith and to inherit the kingdom he promised those who love him?"(James 2:5) Most people on this earth suffer, work hard to buy food and are not living the life of a rich man. It would be impossible for them to identify with and believe in someone born into a wealthy class; that would have disenfranchised most of the world. Moreover, the hierarchy and ruling classes of this world typically leave God out; they tend to glorify themselves with pride, arrogance and all manner of self-righteousness, and are therefore unteachable. Only the humble and those who have suffered know how it is to feel worthless, depressed and worried; therefore they are teachable to God because they really need Him and want Him to give hope and guidance. "He pled the cause of the afflicted and needy, and so it went well with him. Is this not what it means to know me declares the Lord." (Jeremiah 22:16)

"I thank thee, O Father, Lord of heaven and earth, that thou hast hid these things from the wise and prudent, and hast revealed them unto babes..." (Matthew 11:25) Jesus is referring to innocence and humility when he says "Truly I say to you, unless you are converted and become like children, you will not enter the kingdom of heaven. Whoever then humbles himself as this child, he is the greatest in the kingdom of heaven" (Matthew 18:2-4)

Some biblical historians who look at the astronomical science charts back at the time of Jesus's birth conclude that he was born in 6 B.C., on April 17th. This is when Jupiter was in Aries, very bright and large and moved from east to west across the sky for several months. The Magi, or three wise men, distinguished foreigners who were actually scientist- astrologers followed Jupiter all the way from Persia to Jerusalem. The Greek word "star" actually means "radiance." Also notable that Aries was the symbol of King Herod's kingdom. The story of the Magi is found only in Matthew 2:1-12.

When we regard a God who is omniscient everything happens for his rationale. God knew his plan would succeed in Mary and designed her destiny to bring forth his son Jesus. "Thou didst weave me in my mother's womb. When I was made in secret, and skillfully wrought in the depths of the earth." (Psalm 139:13-15) God transcended spirit to flesh and blood, then back again, for he was certain to fulfill what his prophets had predicted. "All that is written must be done." (Luke 24:44) Zechariah, a Levite born in Babylon, finds himself in the middle of Jesus's incarnation, as his name translates in Hebrew to "God remembers to bless at the appointed time."

1. *Rosie Tanabe. "Parthenogenesis." New World Encyclopedia. March 26, 2015*

Chapter 9

In Defense of Women in the Church

Something needs to be said about an issue I have witnessed time and again in several locations, and that is the widespread disenfranchisement felt by young ladies toward the church due to the writings ascribed to Paul in the New Testament. To tell females "Women should remain silent in the churches" (1 Corinthians 14:34) when heard registers oppressive. To deny anyone a voice in a free society is cruel. Freedom of speech is guaranteed in the First Amendment. The question is if Paul did actually write that while he was in Ephesus, as attested by Pope Clement 1 of Rome, circa 96 A.D. The other statement Paul made regarding the restriction of females is "I do not allow a woman to teach or to act superior to a man, but to be quiet." (1 Timothy 2:12). Relative to the teaching field, most teachers in public and private schools are females, so this rule would not work in present times. Incidentally, theologians of today that hold the deutero-Pauline hypothesis have expressed that Paul probably did not write the pastorals (1 Timothy, 2 Timothy and Titus); 1. instead they seem to have been written by Catholic editors after Paul's death, with intentions to imitate what Paul would have said; howbeit this resulted in the establishment of a strictly male hierarchical ministry. Pertaining to this particular syllogism, when we use deductive logic this problem stems from the latter (1 Tim 2:12) as an imitation based upon a misperception of the former, (1 Corinthians 14:34) something the Greek philosopher Plato (circa 450 B.C.) would say is twice removed from the truth.

There are volumes written on the problems of authorship of Paul's epistles; several of the thirteen were likely written by Catholic editors or Paul's scribes Tychicus, Onesimus, or Polycarp of Smyrna, or as some believe Paul's mentor Luke as possibilities. 2. The explanation given by modern theologians to soften the harshness of the 1 Corinthians 14:34 passage is that Paul was addressing a specific, pompous group of wealthy women who were being continually disruptive during his sermon. 3. Since the church was experiencing much fragmentation after Christ's ascension years before this incident and had a range of various church factions including Docetism, Montanism, Gnosticism, Nicolaicism, and the use of mediums to call the dead and other pagan practices not in line with Christianity, Paul chastised these particular loud and divisive women present at his sermon. Unfortunately, this event is mostly misunderstood and used by the church patriarchy to disavow all women for all time. 4. The most important point to be made here is that Paul's intention could not have been to silence woman in the church, because he had commended Phoebe, the Deaconess of Cenchrea near Corinth to the church in Rome. Phoebe was the head of her church, as Paul had called her a "diakonos," which means in Greek "to minister." (Romans 16:1-2) He was jailed in Judea circa 58-60 A.D., the place where previously as Saul he had done his worst killing of Christian men and women. Perhaps in a penitent spirit he wrote Romans to the Jews who had converted to Christianity, but since he was in jail Phoebe was his emissary and physically took the Romans book for presentation to the Roman church in circa 58 A.D.

Additionally, Paul physically preached with Priscilla of Ephesus, who was a minister alongside her minister husband Aquila. (2 Timothy 4:19) Then there is Junia, whom Paul introduces as an apostle who worshipped Christ even before himself. (Romans 16:7) So he would be a hypocrite in the extreme to be purporting one thing while doing the opposite. Incidentally, several other females were church leaders, including Lydia, Deaconess of Thyatira (Acts 16:11-15, 40), and the one that ascended to the papacy, Bishop Theodora of the Saint Praxedis church in 820 A.D. As the historical fragmentation that developed in the very early Christian community had doctrinal disagreements, a minority group of Christians were incorporated into the emerging Catholic (which in Greek means "universal") church and conformed to the political authority thereof. Only the Christian Gnostics preserved the tradition of Mary Magdalene as the "apostle to the apostles," while in juxtaposition the Roman Catholics systematically suppressed and restricted the role of women in the church.

Oppression is like leaven in a church: it's evil hidden inside the good, corrupting from within. A wolf in sheep's clothing, the natural man cannot see it until it results in a loss. Jesus said "Watch out and beware of the leaven of the Pharisees and the Sadducees." (Matthew 16:6) "The Lord is a refuge for the oppressed, a stronghold in times of trouble." (Psalm 9:9)

Paul was called Saul up until his mid-thirties, a Pharisee terrorist responsible for the murder of countless Christians before he converted to Christianity. The highlight of his wicked deeds was his participation in the stoning of Saint Stephen, a Christian Deacon in Jerusalem, circa 36 A.D. (Acts 7:54) Paul was afflicted with conjunctivitis which he called "a thorn in the flesh." (2 Corinthians 12:7 and (Galatians 4:13) Incidentally, Saul was twenty eight when Jesus ascended. When Saul converted to Christianity he went to Asia for three years (today known as Turkey), grew a beard and returned under the name of Paul (Paullus his middle name). When he returned he was thirty-four. Paul's life span was from 5 A.D. to 67 A.D.

Matthew 22:30 describes us in the next life with a simile: "For in the resurrection they neither marry (as men on earth) nor are given in marriage (as women are given), but are like angels in heaven." This implies a more co-equal existence of males and females in the eternal without a hierarchy of gender.

1. *Polycarp of Smyrna and the Pastorals, in From the Early Christianity (Aus der Fruhzeit des Christentums (Tubingen: J.C.B. Mohr. 1963) 197-252.*

2. *Arthur G. Patzia. The Evangelical Quarterly 52.1 (Jan.-Mar. 1980): 27-42. Deutero-Pauline Hypothesis: An Attempt at Clarification.*

3. *Christians for Biblical Equality International (CBE). Priscilla Papers. "Does 1 Timothy 2 Prohibit Women from Teaching, Leading, and Speaking in the Church?" page 5, Conclusion. 2018*

4. *Kirk Macgregor." 1 Corinthians 14:33b-38 as a Pauline Quotation-Refutation Device." CBE International. Priscilla Papers. January 30, 2018*

Chapter 10
God's Message

God's main message is love, and he frames it within a paradigm of marriage. His design emerges very early, in the simple companionship of Adam and Eve and continues throughout the metaphoric parables of Jesus. "For this cause a man shall leave his father and mother, and shall cleave to his wife; and they shall become one flesh." (Genesis 2 :24) God says he built the woman from Adam's rib to "correspond" as a helper for Adam. (Genesis 2:18, 2:20) Adam and Eve were originally gardeners in paradise who were cast out of the heavenly firmament to earth.

The way God seals you into his love is by covenant, where both parties make a contract, or agreement with each other. As in any relationship (God) one calls, the other (you) responds, you commit then he blesses and saves. It can also work the other way around, where you call first; however you must do the repenting and committing. The theme of marriage is instated again in Hosea 2:19: "And I will betroth you to Me forever; yes, I will betroth you to Me in righteousness and in justice, In lovingkindness and in compassion, and I will betroth you to Me in faithfulness. Then you will know the Lord." In classical Hebrew literature the word betrothal means sanctification, inasmuch as you become set apart for God's purpose. This doesn't mean you have to be a minister; you can be an everyday person who adopts Christian values into your life.

One can see a recurrent theme of acceptance versus rejection pursuant to a person's preparation beginning in the Parable of the Ten Virgins, (Matthew 25:1-13) where five virgins were prepared to meet the bridegroom while the other five were not. The five who ran out of oil had to make a dash to the store to buy more oil while the other five, who had brought refills with them stayed on and were there when Jesus appeared and received them, and so went in with him to the wedding feast. Incidentally, the five who were prepared refused to give their refills to the other five who ran out of oil. The unprepared five returned from the store only to find they were too late and locked out of the feast. They had not been ready and so were rejected while clearly the prepared virgins were accepted for thinking ahead. The warning comes at the end of the parable, "Be on the alert then, for you do not know the day nor the hour." (Matthew 25:13) Jesus arrived without a specific appointment known to them, and that reflects the precariousness of our situation in life.

Surely there are consequences to the choices you make in life. When I see countless homeless people living on the streets I wonder if they lived a reckless life when young, not thinking enough of their future when they would become older, and assumed reparations would be made for them- perhaps like the virgins who got locked out they held a false hope that somehow things would work out for them without having to go the extra mile to compete for success. Or did they simply have a string of hard luck that repeated itself until they ended up with nothing? Even so, there are no guarantees in life: we are born alone and die alone, and it can be really hard to make ends meet, but God offers us a chance to strengthen our spirit a la a partnership with him through Jesus if we choose to know him, and thus become like those who understood what they needed to do to gain entry, and with that entry comes eternal life into a preferred place, not merely this earthly plane where the ghosts are. Obviously the higher in heaven you go the closer you get to God. "In my father's house are many dwelling places" said Jesus. (John 14:2) In Aramaic within the older translations the word for the place you will be is "anova" which means a singular place specifically for you in God's heart. Metaphorically, this is where the bridegroom takes you to consummate the marriage.

The continuation of the joining of the bridegroom with his church is shown in the Parable of the Wedding Feast from the book of Matthew 22:1-14, where several attempts are made by the king to invite the "A" list of guests, but they chose to reject the invitation by ignoring it and worse, some killed the king's messengers. This is where the saying "don't kill the messenger" comes from, which means if you must get mad be mad at the news, but don't kill the innocent person delivering the news. If you can control your emotions and think logically with your mind you're a step ahead of the crowd. Despite this setback the king must go on with his party, so anyone on the highway is then invited as a guest to the feast. The highway guests are gentiles that are now invited to be part of God's kingdom. During this important event the king walks around surveying his guests when he spots an inappropriate man ill prepared and out of place. The king becomes offended at this imitator and rejects him by having him bound and thrown out from the party. The fact that he had to be bound and thrown out shows he would not leave willingly; and so it is with false prophets.

The book of Matthew was written to the Jews, so one can see this parable of invitees to a wedding feast could be a metaphor for their rejection of Jesus, the honorary son of the king who is the person the king is having the party for. Although many Hebrews converted in belief that Jesus was their Messiah, orthodox Jews such as the Pharisees and Sadducees rejected him. Conversely, God rejects the false attendees. The concluding statement presented is "Many are invited, but few are chosen." (Matthew 22:14) To be chosen first you need to be aware of the upcoming feast, then you need to prepare before you finally attend it. The man who thought all he had to do was show up found himself devoid of God. Although the Jewish people are the chosen of God, the "A" list, they still need to accept Jesus as their Messiah.

Another parallel to this story is the Parable of the Wheat versus the Tares. Here Jesus divides people into four categories: those who do not understand, those who purport to understand at first but cowardly fall away at the slightest conflict, those who genuinely want to accept Christ but are distracted and tempted by greed and lies of the world, and finally those who have understanding and resolve to love God, and ultimately produce good fruit. Once again we see a corollary of God sorting out the saved from the unsaved in alignment with the theme of rejection and acceptance. At the end of this parable comes a warning to be aware of a saboteur who sows tares, or weeds within the good field of wheat, unbeknown to the farmer who was asleep when his field was sabotaged. The farmer is then shown the weeds have sprung up among his wheat crop during the growth season. Our senses being limited, we cannot be aware of every bad thing around us. By and by we perceive it if we are able to distinguish good from bad. However, if you never learn right from wrong you will be like the first two groups who don't embrace God into their lives after all.

When the workers offer to weed out the tares the farmer says no, because while they are pulling up the weeds they may accidently destroy the good crop. At that stage of growth the weeds look just like the wheat, so the owner instructs the workers to wait until the harvest, then gather up the tares and burn them, but to "gather the wheat into my barn." (Matthew 13:30) While we are in the flesh and blood we cannot tell the truly faithful from the not so. It is after one passes away when God judges the soul, looking at your life and how you have spent it. Once again we see the recurring theme of acceptance versus rejection.

Sometimes there is nothing we can do about the wicked but have patience and trust in God that he will recompense in the fairness of his judgement. Psalm 37:35 observes "I have seen a wicked, violent man spreading himself like a luxuriant tree in its native soil." Remember we're not in heaven yet; we are here where evil oft times prevails against what is right. Well the wicked can live it up to the fullest extent here, because once they die they have no inheritance of life and are no more. Jesus said "Every plant which my heavenly Father did not plant shall be rooted up." (Matthew 15:13) You are either in the will of God or you are not, no in between.

Being a fence sitter is nearly as bad as choosing the wrong side of the fence: "So I will spit you out of my mouth, because you are only warm and not hot or cold." (Revelation 3:16) Clearly God wants commitment. Life is rather short and you have only so much time to love God if you want to live somewhere nice after this brief, corporeal life.

About the Author

The author is a researcher, an advisor, a statistician, a legislative analyst, a mountain climber, a fencer, an actor, a teacher, a photographer, a program designer for gifted programs, new teacher training, literacy groups and testing outcomes. She has worked as an External Evaluator in the California's Underperfoming Schools Program (IIUSP). She holds an A.A. degree, a B.A. degree and a M.S. degree, and others.

Ms. Raleigh has designed four instruction models for the Los Angeles Unified School District, all copyrighted in 1997, while at Eagle Rock High School. She has won a Certificate of Recognition award from the LAUSD Board for "Survival," the best Senior High Standards-Based Instruction Model, but evidently the most popular model was the one called "Transcendentalism," sparking a modern revival of the American Transcendentalist movement which has its fundamental roots in Emerson, Thoreau and Whitman, all of whose works are used in the model. Furthermore, when these thematic models were used at John Burroughs Middle School, Transcendentalism made such a deep impression on several students that they produced an anthology with the help of Literacy Cadre Coordinator Sheila Lieberman. Other teaching models are "Etymology," that provides a study in word origins, and "Literal and Implied Meanings," that explores societal perceptions of gender. Copies of the models are available upon request sent to the P.O. Box address below.

The author has invented a way to improve reading comprehension, and the research paper has been published by The Educational Resources Information Center (ERIC), the U.S. Department of Education office website. First published as "A Constructivist Technique Which Improves Reading Comprehension," SP 038 687, the paper has been updated into modern terms and is now copyrighted as "Using Emotional Intelligence to Improve Reading Comprehension," in 2016. Copies are available by request to the P.O. Box address below.

While working on the language designs of the California Standards Ms. Raleigh was an Associate Director at the UCLA School Management Program under Director Charlotte Higuchi of Center X.

She has designed four multi-cultural classes in conjunction with the J. Paul Getty Museum, the Huntington Museum and the Japanese American Museum from 1995 through 1999. The Japanese American Museum class was held on March 15, 1997, and included a presentation followed by a book signing by Jeanne Wakatsuki Houston, the author of Farewell to Manzanar.

If the reader has any questions about the availability of contents of this manuscript please write to the author at P.O. Box 331, Cody, Wyoming, 82414